Electric And Organic

By

Ian Beardsley

ISBN: 978-1-67812-524-0

ENTITY BOOKS

Table of Contents

Book 1: Bone

Introduction

In comparing biological life to artificial intelligence, I find that the instance of bone is the most compelling. It is at this point that I suggest biological life can be taken as a mathematical structure actually using physical aspects of biological life such as molar mass, density, and atomic radii as the values of the variables. Life seems to only present itself this way if we compare it to another construct, like artificial intelligence. The other extraordinary thing is that bone, which is the fundamental framework around which life is built (muscles are attached to it, skin wrapped around it, organs embedded in it) is described by the fundamental framework around which mathematics is built, algebra. Thus, in this paper dedicated to bone, we have all the framework of fundamental mathematics following from it (ratios, proportions, completing the square, quadratic equations, the golden ratio). As we progress to the form built on the skeleton (muscle on bone) we proceed to the next layer of mathematics, calculus, differential equations, and vector calculus). For instance muscle action is like a damped harmonic oscillator in that the force on the muscle in moving a load is proportional the distance it contracts, and the solution of such a differential equation makes use of the fundamental framework of algebra, namely its solutions are obtained by finding the exponents of e with algebra's quadratic equation or, equivalently, the factorization of a quadratic, or completing the square. More primary to bone are the amino acids and biological elements such as C, N, O, H, which when compared to the AI elements Si, Ge, P, B, As, Ga, of which aspects such as bone are built, and semiconductor components are built for logic gates in AI, these find their expression in an elegant set of equations I laid out in earlier work (AI Biodesign, 2019) which make use of ratios and proportions, such as the golden ratio (a/b=b/c, a=b+c) which are building blocks to bone's algebra just as in they are building blocks to things like amino acids, DNA (in the case of biological life) and Si, Ge, P, B, As, are building blocks to circuit components in AI. The equations that follow on all layers in terms of molar mass, density, and atomic radii, may be of such parallel construct in the need for function in that molar mass, density, and atomic radii, determine the properties of elements and their ensuing compounds.

Part 1

It would seem there is some possibility that as life goes from the fundamental framework to form, so does the math that describes it. I make no attempt to understand why the composition of life is aesthetically pleasing and how that came about, but I think it is due to the need for function. The structure herein found arose when I was comparing biological life to artificial intelligence. The case for bone was so interesting to me that I decided to proceed to muscle and skin. It is because of this structure only revealing itself when comparing the biological to AI, that I might suggest you can only speak about what life is relative to another construct, like AI. Especially where awareness is concerned if we consider the Turing test.

In my exploration of the connection between biological life and AI the most dynamic component is that of bone. It affords us the opportunity to look at:

Multiplying Binomials

Completing The Square

The Quadratic Formula

Ratios

Proportions

The Golden Ratio

The Square Root of Two

The Harmonic Mean

Density of silicon is Si=2.33 grams per cubic centimeter.

Density of germanium is Ge=5.323 grams per cubic centimeter.

Density of hydroxyapatite is HA=3.00 grams per cubic centimeter.

This is

$$\frac{3}{4}Si + \frac{1}{4}Ge \approx HA \quad \text{where} \quad HA = Ca_5(PO_4)_3OH$$

Where HA is the mineral component of bone, Si is an Al semiconductor material and Ge is an Al semiconductor material. This means

$$\frac{Si}{HA}Si + \left[1 - \frac{Si}{HA}\right]Ge = HA$$

The harmonic mean between Si and Ge is HA,...

$$\frac{2SiGe}{Si + Ge} \approx HA$$

This is the sextic,...

$$x^2(x + y)^4 - xy(x + y)^4 + 2xy^2(x + y)^3 - 4x^2y^2(x + y)^2 = 0$$

Which has a solution

$$\frac{Si}{Ge} = \frac{1}{\sqrt{2} + 1}$$

Where x=Si, and y=Ge. It can be solved with the online Wolfram Alpha computational engine. But,...

$$\frac{1}{HA^2}Si^2 - \frac{Ge}{HA^2}Si + \left[\frac{Ge}{HA} - 1\right] = 0$$

$$Si = \frac{1}{2}\left[Ge \pm HA\sqrt{\frac{Ge}{HA^2} - \frac{4Ge}{HA} + 4}\right]$$

$$Si = Ge - HA$$

$$\frac{Si}{HA}Si + \left[1 - \frac{Si}{HA}\right]Ge = HA$$

$$\frac{Si^2}{HA} + Ge - \frac{Si}{HA}Ge \approx HA$$

$$\frac{1}{HA}Si^2 - \frac{Ge}{HA}Si + Ge \approx HA$$

$$\frac{1}{HA^2}Si^2 - \frac{Ge}{HA^2}Si + \frac{Ge}{HA} \approx 1$$

$$\frac{1}{HA^2}Si^2 - \frac{Ge}{HA^2}Si + \frac{Ge}{HA} - 1 \approx 0$$

$$\frac{1}{HA^2}Si^2 - \frac{Ge}{HA^2}Si + \left[\frac{Ge}{HA} - 1\right] = 0$$

$$(x + a)(x + a) = x^2 + 2ax + a^2$$

$$(x + a)^2 = x^2 + 2ax + a^2$$

We see that the square of the binomial is a quadratic where the third term is the square of one half the middle coefficient. This gives us a method to solve quadratics called completing the square:

$$ax^2 + bx + c = 0$$

$$ax^2 + bx = -c$$

$$x^2 + \frac{b}{a}x = -\frac{c}{a}$$

$$\left(\frac{1}{2}\frac{b}{a}\right)^2 = \frac{1}{4}\frac{b^2}{a^2}$$

$$x^2 + \frac{b}{a}x + \frac{1}{4}\frac{b^2}{a^2} = -\frac{c}{a} + \frac{1}{4}\frac{b^2}{a^2}$$

$$\left(x + \frac{1}{2}\frac{b}{a}\right)^2 = \frac{b^2 - 4ac}{4a^2}$$

$$x + \frac{b}{2a} = \pm\frac{\sqrt{b^2 - 4ac}}{2a}$$

$$x = \frac{-b \pm \sqrt{b^2 - 4ac}}{2a}$$

$$\frac{1}{HA^2}Si^2 - \frac{Ge}{HA^2}Si + \left[\frac{Ge}{HA} - 1\right] = 0$$

$$x = \frac{-b \pm \sqrt{b^2 - 4ac}}{2a}$$

$$a = \frac{a}{HA^2} \quad b = -\frac{Ge}{HA^2} \quad c = \left[\frac{Ge}{HA} - 1\right]$$

$$b^2 - 4ac = \frac{Ge^2}{HA^4} - 4\frac{1}{HA^2}\left[\frac{Ge}{HA} - 1\right]$$

$$= \frac{Ge^2}{HA^4} - \frac{4Ge}{HA^3} + \frac{4}{HA^2}$$

$$= \frac{1}{HA^2}\left[\frac{Ge^2}{HA^2} - \frac{4Ge}{HA} + 4\right]$$

$$\sqrt{b^2 - 4ac} = \frac{1}{HA}\sqrt{\left(\frac{Ge}{HA} - 2\right)^2}$$

$$x = \frac{\frac{Ge}{HA^2} \pm \frac{1}{HA}\left[\frac{Ge}{HA} - 2\right]}{\frac{2}{HA^2}}$$

$$= \frac{1}{2}Ge \pm \frac{1}{2}HA\left[\frac{Ge}{HA} - 2\right]$$

$$= \frac{1}{2}Ge \pm \frac{1}{2}Ge - HA$$

$$Si = \frac{1}{2}Ge + \frac{1}{2}Ge - HA$$

$$Si = Ge - HA$$

$$Si \approx Ge - HA$$

$$HA \approx \frac{2SiGe}{Si + Ge}$$

$$Si \approx Ge - \frac{2SiGe}{Si + Ge}$$

$$\frac{(Si + Ge)Ge}{Si + Ge} - \frac{(Si + Ge)Si}{Si + Ge} - \frac{2SiGe}{Si + Ge} = 0$$

$$\frac{Ge^2 - 2SiGe - Si^2}{Si + Ge} = 0$$

$$x^2 - 2xy - y^2 = 0$$

$$x^2 - 2xy = y^2$$

$$x^2 - 2xy + y^2 = 2y^2$$

$$(x - y)^2 = 2y^2$$

$$x - y = \pm\sqrt{2}y$$

$$x = y + \sqrt{2}y$$

$$x = y(1 + \sqrt{2})$$

$$\frac{x}{y} = 1 + \sqrt{2}$$

$$\frac{y}{x} = \frac{1}{\sqrt{2} + 1}$$

$$\frac{Si}{Ge} \approx \frac{1}{\sqrt{2} + 1}$$

A ratio is $\dfrac{a}{b}$ and a proportion is $\dfrac{a}{b} = \dfrac{b}{c}$ which means a is to b as b is to c.

The Golden Ratio (Φ)

$$\frac{a}{b} = \frac{b}{c} \quad \text{and.} \quad a = b + c$$

$$ac = b^2 \quad \text{or} \quad c = \frac{b^2}{a}$$

$$a = b + \frac{b^2}{a}$$

$$\frac{b^2}{a} - a + b = 0$$

$$\frac{b^2}{a^2} - 1 + \frac{b}{a} = 0$$

$$\left(\frac{b}{a}\right)^2 + \frac{b}{a} - 1 = 0$$

$$\left(\frac{b}{a}\right)^2 + \frac{b}{a} + \frac{1}{4} = 1 + \frac{1}{4}$$

$$\left(\frac{b}{a} + \frac{1}{2}\right)^2 = \frac{5}{4}$$

$$\frac{b}{a} = -\frac{1}{2} \pm \frac{\sqrt{5}}{2} \qquad \frac{b}{a} = \frac{\sqrt{5} - 1}{2} \qquad \frac{a}{b} = \frac{\sqrt{5} + 1}{2}$$

$$\phi = \frac{\sqrt{5} - 1}{2} \qquad \Phi = \frac{\sqrt{5} + 1}{2} \qquad \phi = \frac{1}{\Phi}$$

The mineral component of bone hydroxyapatite (HA) is

$$Ca_5(PO_4)_3OH = 502.32\frac{g}{mol}$$

The organic component of bone is collagen which is

$$C_{57}H_{91}N_{19}O_{16} = 1298.67\frac{g}{mol}$$

We have

$$\frac{Ca_5(PO_4)_3OH}{C_{57}H_{91}N_{19}O_{16}} = 0.386795722$$

$$\phi = 0.618033989$$

$$1 - \phi = 0.381966011$$

$$\frac{Ca_5(PO_4)_3OH}{C_{57}H_{91}N_{19}O_{16}} \approx (1 - \phi)$$

$$\frac{0.381966011}{0.386795722}100 = 98.75\%$$

$$\frac{Si}{Ge} = \frac{28.09}{72.61} = 0.386861314 \approx (1 - \phi)$$

$$\frac{Si}{Ge} \approx \frac{Ca_5(PO_4)_3OH}{C_{57}H_{91}N_{19}O_{16}}$$

Part 2

In comparing biological life to artificial intelligence I found the case for bone very compelling for life as a mathematical construct. It is really quite interesting, if not aesthetically pleasing in its structure. I covered that in my paper *Life As A Mathematical Construct*, here I revisit bone using those findings in terms of the intermembral index which compares the forelimbs of vertebrates to their hindlimbs. A ratio greater than one means the forelimbs are longer than the hindlimbs and less than one the hindlimbs are longer. It is this ratio that tells paleontologists a great deal about the manner of propulsion of a vertebrate.

I believe it is best to compare biological life to another construct, like artificial intelligence (AI) That is, intermembral index to silicon (si) and germanium (Ge).

Chimpanzee index is 106 or 1.06 in otherwords as a fraction, meaning their forelimbs are longer than their hind limbs compared to humans, which are around 68-70 or 0.68 to 0.70 meaning their hindlimbs are longer than their forlimbs. Thus we see have their

CHIMPANZEE

forelimbs longer for climbing, arm hanging and swinging activities. The longer hindlimbs of humans mean they depend soley on these for propulsion in bipedal walking. Lucy, the 3.2 million year old hominid (Australopithecus Afarensis) has index 88 intermediate between humans and chimpanzees (0.88) and this due to a shortened humerous, not elongated thigh, showing arm length reduced first in the evolutionary trend toward being bipedal. she probably used hindlimbs for bipedal propulsion and forelimbs for climbing. humerous + radius = 22" and femur + tibia = 32" intermembral index $(i) = 22/32 = 0.6875$

$$\frac{1}{i} \approx \frac{1}{\sqrt{2}+1} + 1 = \sqrt{2} \qquad \frac{Si}{Ge} \approx \frac{1}{\sqrt{2}+1}$$

22"

32"

HUMAN

LUCY

We begin by looking at hydroxyapatite (HA) the mineral component of bone and comparing it to silicon (Si) and germanium (Ge) the most basic components of artificial intelligence (AI).

Density of silicon is Si=2.33 grams per cubic centimeter.

Density of germanium is Ge=5.323 grams per cubic centimeter.

Density of hydroxyapatite is HA=3.00 grams per cubic centimeter.

This is

$$\frac{3}{4}Si + \frac{1}{4}Ge \approx HA \quad \text{where} \quad HA = Ca_5(PO_4)_3OH$$

This gives

$$\frac{Si}{HA}Si + \left[1 - \frac{Si}{HA}\right]Ge = HA \quad \text{and.} \quad \frac{2SiGe}{Si + Ge} \approx HA$$

Which is the sextic

$$x^2(x + y)^4 - xy(x + y)^4 + 2xy^2(x + y)^3 - 4x^2y^2(x + y)^2 = 0$$

Which has solution

$$\frac{Si}{Ge} = \frac{1}{\sqrt{2} + 1}$$

Because if y=Si, and x=Ge, we have

$$Si \approx Ge - HA$$

$$HA \approx \frac{2SiGe}{Si + Ge}$$

$$\frac{Ge^2 - 2SiGe - Si^2}{Si + Ge} = 0$$

$$x^2 - 2xy - y^2 = 0$$

$$\frac{y}{x} = \frac{1}{\sqrt{2} + 1}$$

Thus we see that if i is the intermembral index and Si and Ge subscripted with rho are the densities of silicon and germanium, respectively. Thus,...

$i \approx 0.7$

$$\frac{Si_\rho}{Ge_\rho} \approx \frac{1}{\sqrt{2}+1}$$

And,...

$$\frac{1}{i} \approx \frac{1}{\sqrt{2}+1} + 1$$

Now we introduce the organic component of bone, collagen:

$$Ca_5(PO_4)_3OH = 502.32\frac{g}{mol}$$

$$C_{57}H_{91}N_{19}O_{16} = 1298.67\frac{g}{mol}$$

We have

$$\frac{Ca_5(PO_4)_3OH}{C_{57}H_{91}N_{19}O_{16}} = 0.386795722$$

$$\phi = 0.618033989$$

$$1 - \phi = 0.381966011$$

$$\frac{Ca_5(PO_4)_3OH}{C_{57}H_{91}N_{19}O_{16}} \approx (1 - \phi)$$

$$\frac{0.381966011}{0.386795722}100 = 98.75\%$$

$$\frac{Si}{Ge} = \frac{28.09}{72.61} = 0.386861314 \approx (1 - \phi)$$

$$\frac{Si}{Ge} \approx \frac{Ca_5(PO_4)_3OH}{C_{57}H_{91}N_{19}O_{16}}$$

By molar mass. Where, ϕ is the golden ratio conjugate, is recurrent throughout ratios in vertabrates determined by bone lengths.

Part 3

I find 1/4 coupled with 3/4 is recurrent throughout Nature and that where there is (1/4, 3/4) there is (1/3, 2/3) where 1/3 = 1- 2/3. An example of this would be in Mendelian genetics. 3/4 of the off spring express a recessive trait. This is because while the male has XY sex chromosomes the female has XX sex chromosomes. Thus,…

XY + XX = 3X + Y or 1Y: 3X=>1/3 and 1-1/3=2/3

But we can say,…

XY + XX = X + X + X + Y = 4 elements

And,…

$$\frac{Y}{X+X+X+Y} = \frac{1}{4} elements$$

And, 1-1/4=3/4.

This is how (1/4, 3/4) is coupled with (1/3, 2/3)

Air is about 25% oxygen gas (O2) by volume and 75% nitrogen gas (N2) by volume meaning the molar mass of air as a mixture is:

$$0.25O_2 + 0.75N_2 \approx air$$

By density we find

$$0.25Ge + 0.75Si \approx HA$$

(Density of Ge is 5.323 g/cm^3 and of Si is 2.33 g/cm^3)

$HA = hydroxyapatite = Ca_5(PO_4)_3OH$ is the mineral component of bone.

This is the source of our equations:

$$\frac{Si}{HA}(Si) + \left[1 - \frac{Si}{HA}\right](Ge) \approx HA$$

$$\frac{2(Si)(Ge)}{Si + Ge} \approx HA$$

The density of HA is about 3.00 g per cubic centimeter.

$$\frac{Si}{HA}(Si) + \left[1 - \frac{Si}{HA}\right](Ge) \approx HA$$

$$\frac{1}{HA^2}Si^2 - \frac{Ge}{HA^2}Si + \left[\frac{Ge}{HA} - 1\right] = 0$$

$$Si = \frac{1}{2}\left[Ge \pm HA\sqrt{\frac{Ge}{HA^2} - \frac{4Ge}{HA} + 4}\right]$$

$$Si \approx Ge - HA$$

Thus if we consider

$$\frac{1}{4}O_2 + \frac{3}{4}N_2 \approx air$$

Then we should consider sea water, or ocean:

Thus we have,...

$$NaCl = 58.44 g/mol\ and\ H_2O = 18.02 g/mol$$

$$\frac{1}{3} NaCl \approx H_2O\ and\ \frac{2}{3} H_2O \approx Carbon(C)\ 12.01 g/mol$$

$$\frac{2}{3}\frac{H_2O + NaCl}{NaCl} + \frac{1}{3}\frac{H_2O}{NaCl} \approx 1$$

The Ancient Greeks categorized Nature by 4 elements they called Earth, Air, Water, and Fire. In a sense this was a way of saying solid, gas, liquid, energy. The Earth crust is highest in oxygen, mainly because a lot of that is in water (hydrogen combined with oxygen). We have

Oxygen (O) = 467,100 ppm

Silicon (Si) = 276,900 ppm

Aluminum (Al) = 80,700 ppm

Iron (Fe) = 50,500 ppm

If we say metal is Al +Fe = 131,200 and dirt is 276,900 then metal+dirt=408,100.

Dirt= 276,900/408,100 = 67.85%

Metal = 131200/408,100 = 32%

Thus,...

0.67(dirt) + 0.33(metal) = earth or

$$\frac{1}{3} metal + \frac{2}{3} dirt \approx earth$$

Is (1/3, 2/3) by parts.

$$\frac{1}{4} O_2 + \frac{3}{4} N_2 \approx air$$

Is (1/4, 3/4) by molar mass.

$$\frac{2}{3}\frac{H_2O + NaCl}{NaCl} + \frac{1}{3}\frac{H_2O}{NaCl} \approx 1$$

For Sea or ocean water by molar mass.

$$\frac{1}{4} Ge + \frac{3}{4} Si = bone$$

By density, where Ge and Si are mineral and sand components.

Why notice that the density of Si divided by the density of HA is about 3/4 and write

$$\frac{Si}{HA}Si + \left[1 - \frac{Si}{HA}\right]Ge = HA$$

And then use that with

$$\frac{2SiGe}{Si + Ge} \approx HA$$

When it means we have to solve the sextic

$$x^2(x + y)^4 - xy(x + y)^4 + 2xy^2(x + y)^3 - 4x^2y^2(x + y)^2 = 0$$

When we could have just noticed

$$Si = Ge - HA$$

And then merely have to solve the quadratic

$$x^2 - 2xy - y^2 = 0$$

The reason is that by first looking at the ratios of our densities instead of their differences we have bone as part of the Ancient Greek elements and that takes us to archaeology. What is Archaeology but the study of the decomposition of bone in its exposure to earth, air, and water?

I found it interesting that bone was appearing in the context of Earth, Air, and Water. I knew that the oldest skeletons of humans and their ancient ancestors were unearthed in Africa, and that Chinese archaeologists wanted to find skeletons in China just as old so they could show they arose independently of any other people. The argument as to why they had not found anything as old as in Africa was that Africa is drier so skeletons preserve longer. And, this brings me to my point:

The organic component of bone is collagen and decomposes early on. What is left is the mineral component (hydroxylapatite, HA). The amount of bone in eight weeks that decays in soil (Earth) is the same as the amount of bone that decays in air in two weeks (CAP, 1986) and the primary factors in bone decomposition are soil, air, and water. Mendelian genetics are are at the core of biological evolution and the study of that is the study of bone decomposition. I need only present now, my earlier work. The mineral component (HA) decomposes in exposure to water into calcium and phosphates.

Thus in humid conditions bones can decompose in decades and in dry conditions can be preserved for thousands of years. Francesco Berna, Alan Mathews, and Stephen Weiner report (2004):

We measured the ionic activity products at "steady-state" conditions and we identify a recrystallization window between pH 7.6 and 8.1, which defines the conditions under which bone crystals dissolve and reprecipitate as a more insoluble form of carbonated hydroxyl apatite. As these conditions are common in nature, most fossil bones will not maintain their original crystals with time.

Part 4

We have said

$$\frac{1}{i} \approx \frac{1}{\sqrt{2}+1} + 1 \approx \sqrt{2}$$

$$\frac{Si_\rho}{Ge_\rho} \approx \frac{1}{\sqrt{2}+1}$$

So,...

$$\frac{1}{i} \approx \frac{Si_\rho}{Ge_\rho} + 1 \quad \text{Or...}$$

$$\frac{1}{i} \approx \frac{Si_\rho + Ge_\rho}{Ge_\rho}$$

$$i \approx \frac{Ge_\rho}{Si_\rho + Ge_\rho}$$

We have also said

$$\frac{Si_M}{Ge_M} \approx \frac{Ca_5(PO_4)_3OH}{C_{57}H_{91}N_{19}O_{16}}$$

$$\frac{Ge_\rho}{Si_\rho + Ge_\rho} \approx \frac{\sqrt{2}}{2} = \sin\frac{\pi}{4} = \cos\frac{\pi}{4}$$

$$\frac{Si_M}{Ge_M} \approx (1-\phi)$$

$$Si_M \approx (1-\phi)Ge_M$$

$$Si_M \approx \frac{Ca_5(PO_4)_3OH}{C_{57}H_{91}N_{19}O_{16}}Ge_M$$

In order for Si and Ge to semiconductor they must be doped. Often the doping agents are phosphorus (P) and boron (B). We find by atomic radius where SI_R=110pm, Ge_R=125pm, P_R=100pm and B_r=85pm

$$\frac{P_R + B_R}{Si_R} \approx \Phi \quad \text{or,....} \quad \frac{Si_R}{P_R + B_R} \approx \phi$$

But if we take the arithmetic mean between the geometric mean of P_M and B_M and the harmonic mean and divide by Si_M we find that,...

$$\frac{\sqrt{P_M B_M}(P_M + B_M) + 2P_M B_M}{2(P_M + B_M)Si_M} \approx \phi$$

Which yields,...

$$\frac{2Si_R}{P_R + B_R}Si_M \approx \sqrt{P_M P_M} + \frac{2P_M B_M}{P_M + B_M}$$

Thus,...

$$\frac{2Si_R}{P_R + B_R} \approx \frac{C_{57}H_{91}N_{19}O_{16}}{Ca_5(PO_4)_3OH}\left[\sqrt{P_M B_M} + \frac{2P_M B_M}{P_M + B_M}\right]\frac{1}{Ge_M}$$

$$\frac{Ge_\rho}{Si_\rho + Ge_\rho} \approx \frac{\sqrt{2}}{2}$$

Thus since we have

$$\frac{2Si_R}{P_R + B_R} \approx \frac{C_{57}H_{91}N_{19}O_{16}}{Ca_5(PO_4)_3OH}\left[\sqrt{P_M B_M} + \frac{2P_M B_M}{P_M + B_M}\right]\frac{1}{Ge_M}$$

$$\frac{Ge_p}{Si_p + Ge_p} \approx \frac{\sqrt{2}}{2}$$

And, since we have,...

$$CO_2 = 12.01 + 32.00 = 44.01$$

$$O_2 = 32.00$$

$$\frac{CO_2}{O_2} = \frac{44.01}{32.00} = 1.375 \approx 1.414 = \sqrt{2}$$

Then we can write as well

$$\frac{2Si_R}{P_R + B_R} \approx \frac{C_{57}H_{91}N_{19}O_{16}}{Ca_5(PO_4)_3OH}\left[\sqrt{P_M B_M} + \frac{2P_M B_M}{P_M + B_M}\right]\frac{1}{Ge_M}$$

$$\frac{2Ge_p}{Si_p + Ge_p} \approx \frac{CO_2}{O_2}$$

CO2 (carbon dioxide) is what animal life exhales when it breathes, which is taken up by plant life to make the oxygen gas (O2) that it breathes in a process called photosynthesis where plants get energy from the sun to make the monomer used to make its food the carbohydrate CH2O at the basis of making the more complex sugars at the bottom of the food chain. The reaction is:

$$CO_2 + 2H_2O + photons \longrightarrow CH_2O + O_2 + H_2O$$

Part 5

We look at the fourth Greek element, Fire. We choose peat as our tinder as it is naturally occurring and would have been available to ancient humans for making fire. It has one of the lowest auto ignition temperatures, which is

$227 \, °_C$

Peat has a specific heat of

$1.88 kJ/kg \, °_K$ Or,... $0.45 kcal/kg \, °_C$

$1 kcal = 4148 J$

$$\frac{0.45 kcal}{kg \, °_C} \frac{4184 J}{kcal} \frac{kg}{1000g} = 0.0018828 J/g \, °_C$$

$$\frac{0.0018828 J}{g \, °_C} \frac{1 \times 10^7 ergs}{J} = 18828 ergs/g \, °_C$$

$$\frac{g \, °_C}{0.0018828 ergs} \frac{1}{227 \, °_C} = 0.012 (grams \, of \, peat)/erg$$

If we are to say building artificial intelligence (AI) is taking us back to our beginnings of making fire,... then

If we are to have series of switches that are either on or off so we can encode in binary... we need doped silicon or doped germanium (semiconductors like diodes). These have forward biases (current required to turn them on) of

0.6 Volts for silicon and,

0.3 Volts for germanium

The work done in accelerating an electron through a potential of one volt is

$1.6 \times 10^{-19} Joules$

Then there are

$(0.6V)(1.6 \times 10^{-19} J) = 9.6 \times 10^{-20} J$

To turn on a silicon diode. For germanium it is

$(0.3J)(1.6 \times 10^{-19}) = 4.8 \times 10^{-20} J$

These are

$$\frac{(9.6 \times 10^{-20} J)}{1} \frac{(1 \times 10^7 ergs)}{Joule} = 9.6 \times 10^{-13} ergs$$

$$\frac{(4.8 \times 10^{-20} J)}{1} \frac{(1 \times 10^7 ergs)}{Joule} = 4.8 \times 10^{-13} ergs$$

$$\frac{(0.012 grams of peat)}{erg} \frac{(9.6 \times 10^{-13} ergs)}{1} = 1.152 \times 10^{-14} g$$

The density of dry peat is

$$0.4 g/cm^3$$

Thus we have the energy to turn on a silicon diode will burn

$$\frac{1.152 \times 10^{-14} g}{1} \frac{cm^3}{0.4g} = 2.88 \times 10^{-14} cm^3$$

Of peat. If we consider a strand of peat to be approximately cylindrical and to have a length of 12 times its radius then,...

$$V = \pi r^2 h$$

$$2.88 \times 10^{-14} cm^3 = \pi \left(\frac{1}{12} h\right)^2 h = (0.0218) h^3$$

And,...

$$h = 0.0001 cm$$

Thus we say the energy to turn on a silicon diode in an Al circuit is about the energy to burn a ten thousandth of a centimeter of peat, which from experience I think is the amount of tinder we ignite with sparks from hitting together two pieces of flint. Once that small piece is ignited, it ignites the whole piece of tinder and that in turn ignites a pile of tinder used to set fire to kindling.

Thus in looking at the fourth Ancient Greek element fire we return to our original problem of Al.

Part 6

we should consider the densities of
Si and Ge, which are written $Si\rho$ and
$Ge\rho$, and the molar masses of Si and Ge,
which are written SiM and GeM, with
respect to one another:

$$\frac{Si\rho}{Ge\rho} \approx \frac{1}{\sqrt{2}+1}$$

$$\frac{SiM}{GeM} \approx 1 - \phi$$

where $\phi = \frac{\sqrt{5}-1}{2}$

and $\frac{\sqrt{2}}{2} = \sin 45° = \cos 45°$

$\sqrt{2}$ is the ratio of the diagonal
of a square to its sides and is
two right triangles with angles 45°
each:

And ϕ is
formed by
the triangle
with legs 1
and 2:

$$\frac{Si M}{Ge M} \approx \left(\frac{Si\rho}{Ge\rho}\right)$$

$$\frac{1}{\sqrt{2}+1} \cdot \left(\frac{\sqrt{5}-1}{2}\right)$$

The intermembral index, i, fo
humans is $i = \frac{\sqrt{2}}{2} = 0.7$

$$\frac{S_{i\rho}}{Ge_\rho} = \frac{1}{\sqrt{2}+1} \cdot \left[1 - \frac{\sqrt{5}-1}{2}\right] \quad \text{Interestingly}$$

$$2\cos 45° = 2\cos \frac{\pi}{4} = \sqrt{2}$$

$$2\cos 36° = 2\cos \frac{\pi}{5} = \Phi$$

$$\text{where } \Phi = \frac{\sqrt{5}+1}{2} = \frac{1}{\alpha}$$

$$2\cos 30° = 2\cos \frac{\pi}{6} = \sqrt{3}$$

we can show

$$-2\cos\left(\frac{\pi}{4}\right) + 2\cos\left(\frac{\pi}{5}\right) + 2\cos\left(\frac{\pi}{6}\right) \approx \frac{air}{H_2O}$$

by molar mass where $air = 0.25O_2 + 0.75N_2$

$$= 29.0 \text{ g/mol}$$

Thus where $\frac{\sqrt{2}}{2}$ is i in humans

$\frac{\sqrt{5}-1}{2}$ is in humans as well where

the distance from head to navel compared to navel to bottom of the feet is $\emptyset$.

$i = \frac{a}{b}$

$\emptyset = \frac{c}{d}$

Thus we have the very interesting equations

$$\left[1 - \frac{Si_\rho}{Ge_\rho}\right] \approx \frac{(1 - \phi)}{i}$$

$$i = cos\frac{\pi}{4}, \phi = 2cos\frac{\pi}{5} - 1$$

$$\frac{Si_\rho}{Ge_\rho}\frac{Si_M}{Ge_M} \approx \frac{(1 - \phi)}{(2i + 1)}$$

Journal Notes

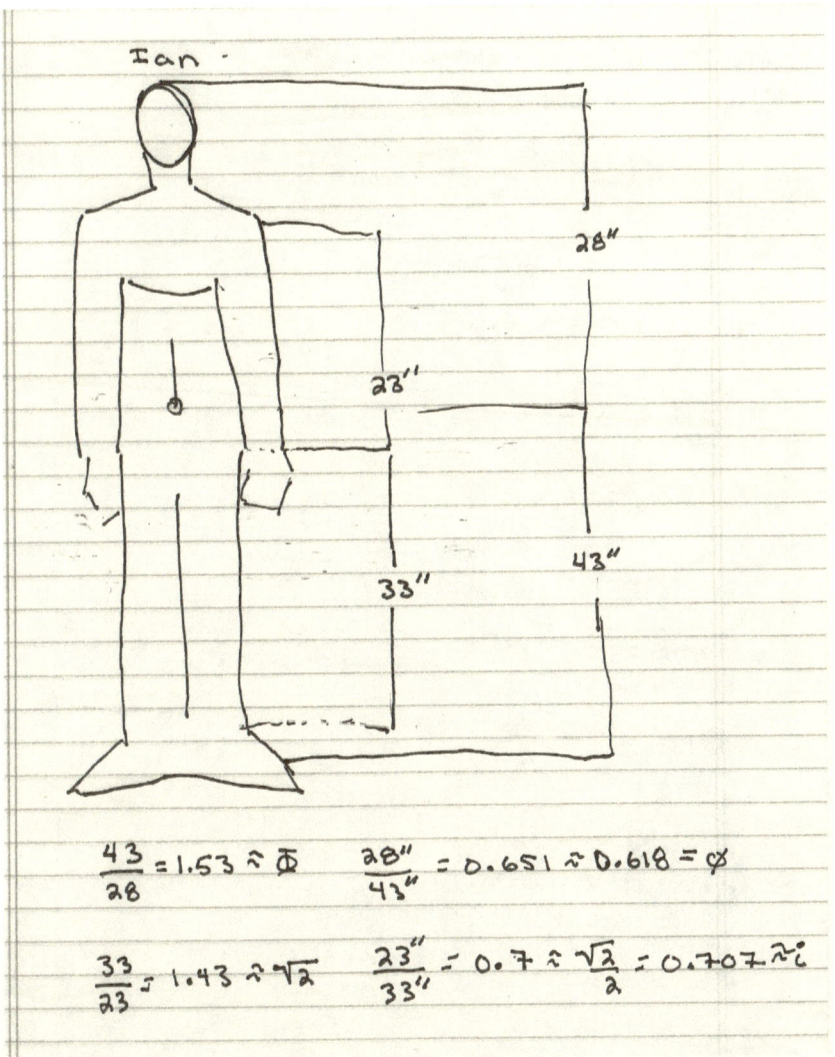

Ian

$$\frac{43}{28} = 1.53 \approx \Phi \qquad \frac{28''}{43''} = 0.651 \approx 0.618 = \phi$$

$$\frac{33}{23} = 1.43 \approx \sqrt{2} \qquad \frac{23''}{33''} = 0.7 \approx \frac{\sqrt{2}}{2} = 0.707 \approx$$

$a + b = 23 + 33 = 56 = A$

$c + d = 28 + 43 = 71 = B$

$\dfrac{A}{B} = 0.7887$ $\dfrac{B}{A} = 1.267857$

$a =$ humerous + radius

$b =$ femur + tibia

$\dfrac{a+b}{c+d} = 0.7887 = \dfrac{56}{71}$

arm = ~~23234~~ $a = $ ~~23~~ 23 ~~234~~

leg = ~~38~~ $b = $ ~~mm~~ 33 mm

humerous = 13"
radius = 10" $10 + 13 = 23$

femur = 17"
tibia = 16" $16 + 17 = 33$

humerous = w $a = w + x$
radius = x

femur = y $b = y + z$
tibia = z

Let's say here what is trying to
be expressed here is that

$$\frac{B}{A} = \frac{5}{4}$$

All of these ratios can be taken
as levers for which there are three
cases, each serving different ends.

1) The fulcrum is between the effort
   and the load for which the mechanical
   advantage may be greater than 1,
   less than 1, or equal to 1. A crow bar.

2) The load is between the effort
   and the fulcrum. The mechanical
   advantage is always greater than 1.
   It is a force multiplier. A wheelbarrow.

3) The effort is between the fulcrum
   and the load. It is a speed multiplier.
   A pair of tweezers, a hammer.

Book 2: Carbon And Silicon

This is the prequel to *Bone*, where I compared the aspect of biological life that is bone, to artificial intelligence (AI). This work deals with the comparison of biological life to artificial intelligence where the elements with which the former are built (CHNOPS) to the elements with which the latter are built (Si, Ge, P, B, Ga, As) are concerned. It is a purpose of biological life (C, N, O, H) to discover the properties of (P, B, Si) so it can make computing machines which are necessary to its survival.

Carbon And Silicon

The golden ratio and the golden ratio conjugate are the solution of the quadratic

$$\left(\frac{a}{b}\right)^2 - \frac{a}{b} - 1 = 0 \text{ that meets the conditions } \frac{a}{b} = \frac{b}{c} \text{ and } a=b+c$$

Where $\Phi = \frac{a}{b}$ and $\phi = \frac{\sqrt{5}-1}{2}$, $\phi = \frac{1}{\Phi}$.

We guess that artificial intelligence (AI) has the golden ratio, or its conjugate in its means geometric, harmonic, and arithmetic by molar mass by taking these means between doping agents phosphorus (P) and boron (B) divided by semiconductor material silicon (Si) :

$$\frac{\sqrt{PB}}{Si} = \frac{\sqrt{(30.97)(10.81)}}{28.09} = 0.65$$

$$\frac{2PB}{P+B}\frac{1}{Si} = \frac{2(30.97)(10.81)}{30.97+10.81}\frac{1}{28.09} = 0.57$$

$$\frac{0.65+0.57}{2} = 0.61 \approx \phi$$

Which can be written

$$\frac{\sqrt{PB}(P+B)+2PB}{2(P+B)Si} \approx \phi$$

We see that the biological elements, H, N, C, O compared to the AI elements P, B, Si is the golden ratio conjugate (phi) as well:

$$\frac{C+N+O+H}{P+B+Si} \approx \phi$$

So we can now establish the connection between artificial intelligence and biological life:

$$(P+B+Si)\frac{\sqrt{PB}(P+B)+2PB}{2(P+B)Si} \approx (C+N+O+H)$$

Which can be written:

$$\sqrt{PB}\left[\frac{P}{Si}+\frac{B}{Si}+1\right]+\frac{2PB}{P+B}\left[\frac{P}{Si}+\frac{B}{Si}+1\right]\approx 2HCNO$$

Where HNCO is isocyanic acid, the most basic organic compound. We write in the arithmetic mean:

$$\left[\sqrt{PB}+\frac{2PB}{P+B}+\frac{P+B}{2}\right]\left[\frac{P}{Si}+\frac{B}{Si}+1\right]\approx 3HNCO$$

Which is nice because we can write in the second first generation semiconductor as well (germanium) and the doping agents gallium (Ga) and arsenic (As):

$$\left[\sqrt{PB}+\frac{2PB}{P+B}+\frac{P+B}{2}\right]\left[\frac{P}{Si}+\frac{B}{Si}+1\right]\approx HNCO\left[\frac{Ga}{Ge}+\frac{As}{Ge}+1\right]$$

Where

$$\frac{Zn}{Se}\approx\frac{\left[\frac{P}{Si}+\frac{B}{Si}+1\right]}{\left[\frac{Ga}{Ge}+\frac{As}{Ge}+1\right]}$$

Where ZnSe is zinc selenide, an intrinsic semiconductor used in AI, meaning it doesn't require doping agents. We now have:

$$\sqrt{PB}\left(\frac{Zn}{Se}\right)+\frac{2PB}{P+B}\left(\frac{Zn}{Se}\right)+\frac{P+B}{2}\left(\frac{Zn}{Se}\right)\approx HNCO$$

We could begin with semiconductor germanium (Ge) and doping agents gallium (Ga) and phosphorus (P) and we get a similar equation:

$$\frac{2GaP}{Ga+P}=42.866,\quad \sqrt{GaP}=46.46749$$

In grams per mole. Then we compare these molar masses to the molar masses of the semiconductor material Ge:

$$\frac{2GaP}{Ga+P}\frac{1}{Ge}=\frac{42.866}{72.61}=0.59$$

$$\sqrt{GaP}\frac{1}{Ge}=\frac{46.46749}{72.61}=0.64$$

Then, take the arithmetic mean between these:

$$\frac{0.59 + 0.64}{2} = 0.615$$

We then notice this is about the golden ratio conjugate, ϕ, which is the inverse of the golden ratio, Φ. $\phi \approx \frac{1}{\Phi}$. Thus, we have

1. $\dfrac{\sqrt{GaP}(Ga + P) + 2GaP}{2(Ga + P)Ge} \approx \phi$

2. $\dfrac{\sqrt{GaP}(Ga + P) + 2GaP}{2(Ga + P)Si} \approx \Phi$

This is considering the elements of artificial intelligence (AI) Ga, P, Ge, Si. Since we want to find the connection of artificial intelligence to biological life, we compare these to the biological elements most abundant by mass carbon (C), hydrogen (H), nitrogen (N), oxygen (O), phosphorus (P), sulfur (S). We write these CHNOPS (C+H+N+O+P+S) and find:

$$\frac{CHNOPS}{Ga + As + Ge} \approx \frac{1}{2}$$

A similar thing can be done with germanium, Ge, and gallium, Ga, and arsenic, As, this time using CHNOPS the most abundant biological elements by mass:

$$\left[\sqrt{GaAs} + \frac{2GaAs}{Ga + As} + \frac{Ga + As}{2}\right]\left[\frac{Ga}{Ge} + \frac{As}{Ge} + 1\right] \approx CHNOPS\left[\frac{Ga}{Si} + \frac{As}{Si} + 1\right]$$

$$\sqrt{GaAs}\left(\frac{O}{S}\right) + \frac{2GaAs}{Ga + As}\left(\frac{O}{S}\right) + \frac{Ga + As}{2}\left(\frac{O}{S}\right) \approx CHNOPS$$

$$\frac{O}{S} \approx \frac{\left[\frac{Ga}{Ge} + \frac{As}{Ge} + 1\right]}{\left[\frac{Ga}{Si} + \frac{As}{Si} + 1\right]}$$

$$\frac{\sqrt{GaAs}(Ga + As) + 2GaAs}{2(Ga + As)Ge} \approx 1$$

$$\frac{C + H + N + O + P + S}{Ga + As + Ge} \approx \frac{1}{2}$$

We can also make a construct for silicon doped with gallium and phosphorus:

$$(C + N + O + H) \approx \frac{2(Ga + P)Si}{\sqrt{GaP}(Ga + P) + 2GaP}(P + B + Si)$$

$$HNCO \approx \frac{2(Ga + P)Si}{(Ga + P)\left[\sqrt{GaP} + \frac{2GaP}{Ga + P}\right]}(P + B + Si)$$

$$HNCO \approx \frac{2(P + B + Si)Si}{\sqrt{GaP} + \frac{2GaP}{Ga + P}}$$

And we have for germanium doped with gallium and phosphorus:

$$\frac{\sqrt{GaP}(Ga + P) + 2GaP}{2(Ga + P)Ge} \approx \phi$$

$$\left[\sqrt{GaP} + \frac{2GaP}{Ga + P} + \frac{Ga + P}{2}\right]\left[\frac{P}{Ge} + \frac{B}{Ge} + \frac{Si}{Ge}\right] \approx HNCO\left[\frac{Ga}{Ge} + \frac{As}{Ge} + 1\right]$$

$$\sqrt{GaP}\left(\frac{B}{S}\right) + \frac{2GaP}{Ga + P}\left(\frac{B}{S}\right) + \frac{Ga + P}{2}\left(\frac{B}{S}\right) \approx HNCO$$

The Fundamental Albioequations

$$\left[\sqrt{PB} + \frac{2PB}{P+B} + \frac{P+B}{2}\right]\left[\frac{P}{Si} + \frac{B}{Si} + 1\right] \approx HNCO\left[\frac{Ga}{Ge} + \frac{As}{Ge} + 1\right]$$

$$\left[\sqrt{GaAs} + \frac{2GaAs}{Ga+As} + \frac{Ga+As}{2}\right]\left[\frac{Ga}{Ge} + \frac{As}{Ge} + 1\right] \approx CHNOPS\left[\frac{Ga}{Si} + \frac{As}{Si} + 1\right]$$

$$\left[\sqrt{GaP} + \frac{2GaP}{Ga+P} + \frac{Ga+P}{2}\right]\left[\frac{P}{Ge} + \frac{B}{Ge} + \frac{Si}{Ge}\right] \approx HNCO\left[\frac{Ga}{Ge} + \frac{As}{Ge} + 1\right]$$

$$HNCO \approx \frac{2(P+B+Si)Si}{\sqrt{GaP} + \frac{2GaP}{Ga+P}}$$

$$\frac{\sqrt{PB}(P+B) + 2PB}{2(P+B)Si} \approx \phi$$

$$\frac{\sqrt{GaAs}(Ga+As) + 2GaAs}{2(Ga+As)Ge} \approx 1$$

$$\frac{\sqrt{GaP}(Ga+P) + 2GaP}{2(Ga+P)Ge} \approx \phi$$

$$\frac{\sqrt{GaP}(Ga+P) + 2GaP}{2(Ga+P)Si} \approx \Phi$$

$$\frac{C+N+O+H}{P+B+Si} \approx \phi$$

$$\frac{C+H+N+O+P+S}{Ga+As+Ge} \approx \frac{1}{2}$$

$$\frac{Zn}{Se} \approx \frac{\left[\frac{P}{Si} + \frac{B}{Si} + 1\right]}{\left[\frac{Ga}{Ge} + \frac{As}{Ge} + 1\right]}$$

$$\frac{O}{S} \approx \frac{\left[\frac{Ga}{Ge} + \frac{As}{Ge} + 1\right]}{\left[\frac{Ga}{Si} + \frac{As}{Si} + 1\right]}$$

We now want to write out the equations for atomic radius, density, and molar mass as these are the components upon which the properties of the elements should rely.

P_R	Radius Phosphorus	100 pm	
B_R	Radius Boron	85 pm	
Si_R	Radius Silicon	110 pm	
Ga_R	Radius Gallium	130 pm	
As_R	Radius Arsenic	115 pm	
Ge_R	Radius Germanium	125 pm	
P_M	Molar Mas Phosphorus	30.97 g/mol	
B_M	Molar Mass Boron	10.81 g/mol	
Si_M	Molar Mass Silicon	28.09 g/mol	

You will find:

$$\left[\frac{Ga_R}{Ge_R} + \frac{As_R}{Ge_R} + 1 \right] \approx \pi \quad \text{and} \quad \left[\frac{P_R}{Si_R} + \frac{B_R}{Si_R} \right] \approx \Phi$$

Or,...

$$\frac{P_R + B_R}{Si_R} \approx \Phi \quad \text{Or,....} \quad \frac{Si_R}{P_R + B_R} \approx \phi$$

We now subscript the elements with M for molar mass and find:

$$\frac{\sqrt{P_M B_M}(P_M + B_M) + 2P_M B_M}{2(P_M + B_M)Si_M} \approx \phi$$

Which can be written,...

$$\frac{\sqrt{P_M B_M} + \frac{2P_M B_M}{(P_M + B_M)}}{2Si_M} \approx \phi$$

Which yields,...

$$\frac{2Si_R}{P_R + B_R} Si_M \approx \sqrt{P_M B_M} + \frac{2P_M B_M}{P_M + B_M}$$

Amino Acids

In order to have biological life we need to have carbon. The Astronomer Fred Hoyle figured out how carbon is made by stars. It starts from the formation of helium from hydrogen then from there helium forms carbon in the triple alpha process:

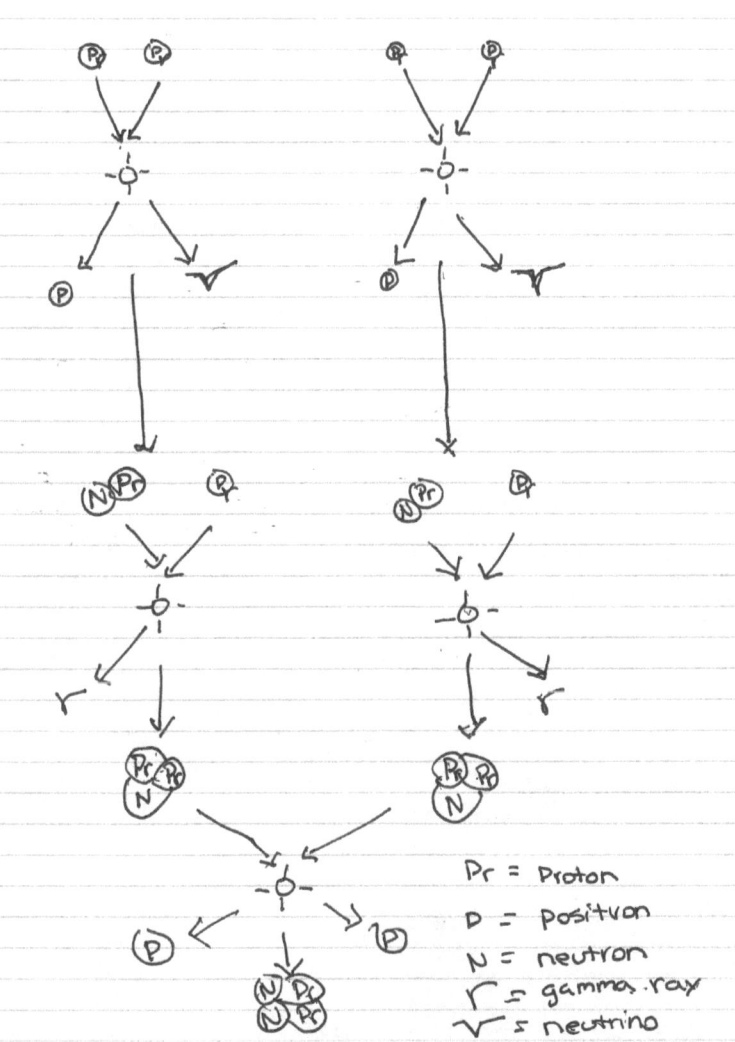

Pr = Proton
D = positvon
N = neutron
Γ = gamma ray
V = neutrino

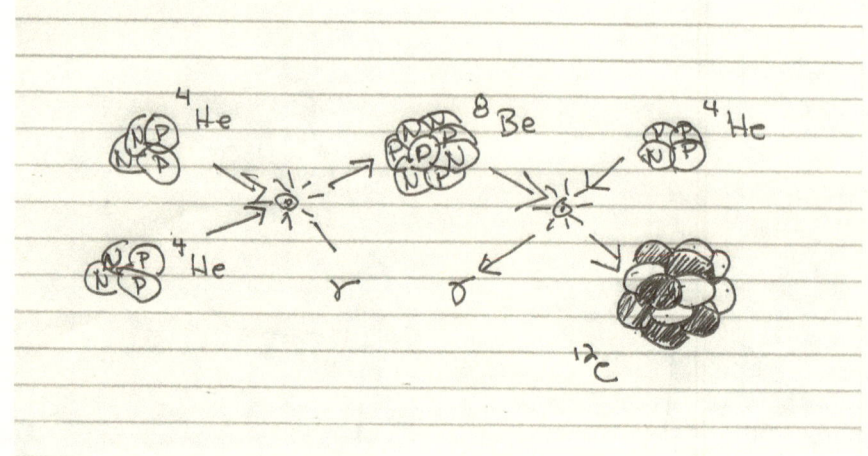

Conducted in 1952 by Stanley Miller and Harold Urey they simulated the primordial conditions thought to exist in the early Earth by mixing Water (H2O) methane (CH4) ammonia (NH3) and hydrogen gas (H2) in a flask and passing a spark through it. The result was the production of 11 of the 20 genetically encoded amino acids. It is thought that the reactions that formed amino acids like this are like a reaction that was recorded a century earlier by Strecker called the strecker synthesis which was:

Aldehyde	Ammonia			Imine	H

$$
\begin{array}{ccccccc}
\text{Aldehyde} & & \text{Ammonia} & & & \text{Imine} & H \\
O & & & & OH & & / \\
\| & & & & \| & & R\text{-}\text{-}C + H_2O \\
R\text{-}\text{-}C\text{-}\text{-}H & + & NH^3 & \xrightarrow{\quad} & R\text{-}C\text{-}H & \xrightarrow{\quad} & \| \\
& & & & \| & & NH \\
& & & & NH_2 & & \\
\end{array}
$$

$$
\begin{array}{ccccccc}
H\ \ O & & & & H\ \ O & & H\ \ \ V \\
|\ \ \| & & H_2O & & |\ \ \| & H_2O & |\ \ \ \\
NH_3+\ R\text{-}C\text{-}C\text{-}\text{-}OH & \xleftarrow{\quad} & & & R\text{-}C\text{-}C\text{-}NH_2 & \xleftarrow{\quad} & R\text{-}C\text{-}C\equiv N \\
|\ \ & & \text{heat} & & |\ \ & & |\ \ \\
NH_2 & & & & NH_2 & & NH_2 \\
\text{amino acid} & & & & & & \text{aminocyanonitril} \\
\end{array}
$$

HCN

I processed the 20 genetically encoded amino acids according to the following scheme:

$$\frac{aminogroup}{acidgroup}(RGroup)$$

In hopes of finding a connection between artificial intelligence and the biological. The result was that two of the amino acids were equal to elements in the periodic table of the elements and they were perfectly carbon (C) the core element of biological life, and silicon (Si) the core element of of artificial intelligence. The amino acids are the building blocks of life, synthesized into proteins by DNA. The two amino acids were serine and glutamine as follows,...

$$\frac{H_3N}{COO}(CH_2 + OH) = C$$

$$\frac{H_3N}{COO}(2CH_2 + CO + NH_2) = Si$$

$$\frac{H_3N}{COO} = (1 - \phi) \qquad \phi = \frac{\sqrt{5} - 1}{2}$$

$$(1 - \phi) = \frac{Si}{Ge}$$

$$\frac{Si}{Ge}(CH_2 + OH) = C$$

$$(CH_2 + OH) = \frac{C}{Si}Ge$$

$$\frac{Si}{Ge}(2CH_2 + CO + NH_2) = Si$$

$$(2CH_2 + CO + NH_2) = Ge$$

Ge is the other core semiconductor element.

$$\frac{(CH_2 + OH)}{(2CH_2 + CO + NH_2)} = \frac{C}{Si}$$

$$C = \frac{(CH_2 + OH)}{(2CH_2 + CO + NH_2)}Si$$

These equation are nearly 100% accurate.

$CH_2+OH=12.01+2(1.01)+16,00+1.01=14.03+17.01=31.04$ g/mol

$Si/Ge=28.09/72.61=0.38686$

$(0.38686)(31.04)=12.008$

$C=12.01$

$$\frac{12.008}{12.01}100 = 99.98\%$$

$2CH_2=2(12.01+2.02)=28.06$

$CO=12.01+16.00=28.01$

$NH_2=14.01+2.02=16.03$

$28.06+28.01+16.03=72.1$

$Ge=72.61$

$$\frac{72.1}{72.61}100 = 99.2976\%$$

The idea is to try to understand biological life, in particular its origins, by looking at something we understand, artificial intelligence.

The equations imply:

$$\frac{H_3N}{COO}Ge \approx Si$$

The primordial compounds from which amino acids are made — water (H_2O) methane (CH_4) and ammonia (NH_3)—seem to be related to primitive AI which would be a tungsten filament (W) encased in a glass tube (SiO_2) to make vacuum tubes for switches as follows:

$$\frac{W}{SiO_2} \approx \frac{H_2O}{CH_4} + \frac{NH_3}{CH_4} + 1$$

DNA

In order to have DNA and RNA we need the sugars deoxyribose and ribose. But sugars are polymers of CH2O which itself is not a sugar but has the same structure as a sugar which is (CH2O)n. When we consider the Miller-Urey experiment where Miller and Urey produced from the hypothesized primordial earth composition of NH3 (ammonia), CH4 (methane) and H2O (water) some of the 20 genetically encoded amino acids that are the building blocks of life, then it is interesting that these are equal to the ancient, "primordial" artificial intelligence which were switches made of tungsten filaments (W) encased in glass (SiO2):

$$\frac{W}{SiO_2} \approx \left[\frac{H_2O}{CH_4} + \frac{NH_3}{CH_4} + 1 \right]$$

Because the monomer from which sugars were surely made as DNA came into existence on the primordial earth has it equivalence in

$$\frac{SiO_2}{CH_2O} = 2$$

$$\frac{\left[\frac{Ga}{Si} + \frac{As}{Si} + 1 \right]}{\left[\frac{Ga}{Ge} + \frac{As}{Ge} + 1 \right]} \approx 2$$

So that,...

$$CH_2O \approx SiO_2 \frac{\left[\frac{Ga}{Si} + \frac{As}{Si} + 1 \right]}{\left[\frac{Ga}{Ge} + \frac{As}{Ge} + 1 \right]}$$

Where Ga and Ge are gallium and arsenic, doping agents for germanium (Ge) and silicon (Si) the first and second generation semiconductors, respectively.

And indeed DNA consists of the sugar deoxyribose, but it is made as well of the nitrogenous bases adenine, guanine, cytosine, and thymine. These are:

Adenine=C5H5N5=135.13 g/mol

Guanine=C5H5N5O=151.13 g/mol

Cytosine=C5H4N3O=111.1 g/mol

Thymine=C5H6N2O2=126.113 g/mol

And we have the sum of their ratios is pi, the ratio of the circumference of a circle to its diameter:

$$\frac{C_5H_5N_5}{C_5H_6N_2O_2} + \frac{C_5H_5N_5O}{C_5H_6N_2O_2} + \frac{C_5H_4N_3O}{C_5H_6N_2O_2} \approx \pi$$

And that,...

$$\frac{C_5H_5N_5}{C_5H_6N_2O_2} + \frac{C_5H_5N_5O}{C_5H_6N_2O_2} + \frac{C_5H_4N_3O}{C_5H_6N_2O_2} \approx \frac{Ga}{Ge} + \frac{As}{Ge} + 1$$

And we have to consider its phosphate backbone, H3PO4

$$\frac{Se}{Zn} \approx \frac{H_3PO_4}{C_5H_6N_2O_2}$$

Where ZnSe is zinc selenide, an intrinsic semiconductor.

DNA is, as well, approximately equal to the primordial prebiotic substances:

$$\frac{C_5H_5N_5}{C_5H_6N_2O_2} + \frac{C_5H_5N_5O}{C_5H_6N_2O_2} + \frac{C_5H_4N_3O}{C_5H_6N_2O_2} \approx \frac{H_2O}{CH_4} + \frac{NH_3}{CH_4} + 1$$

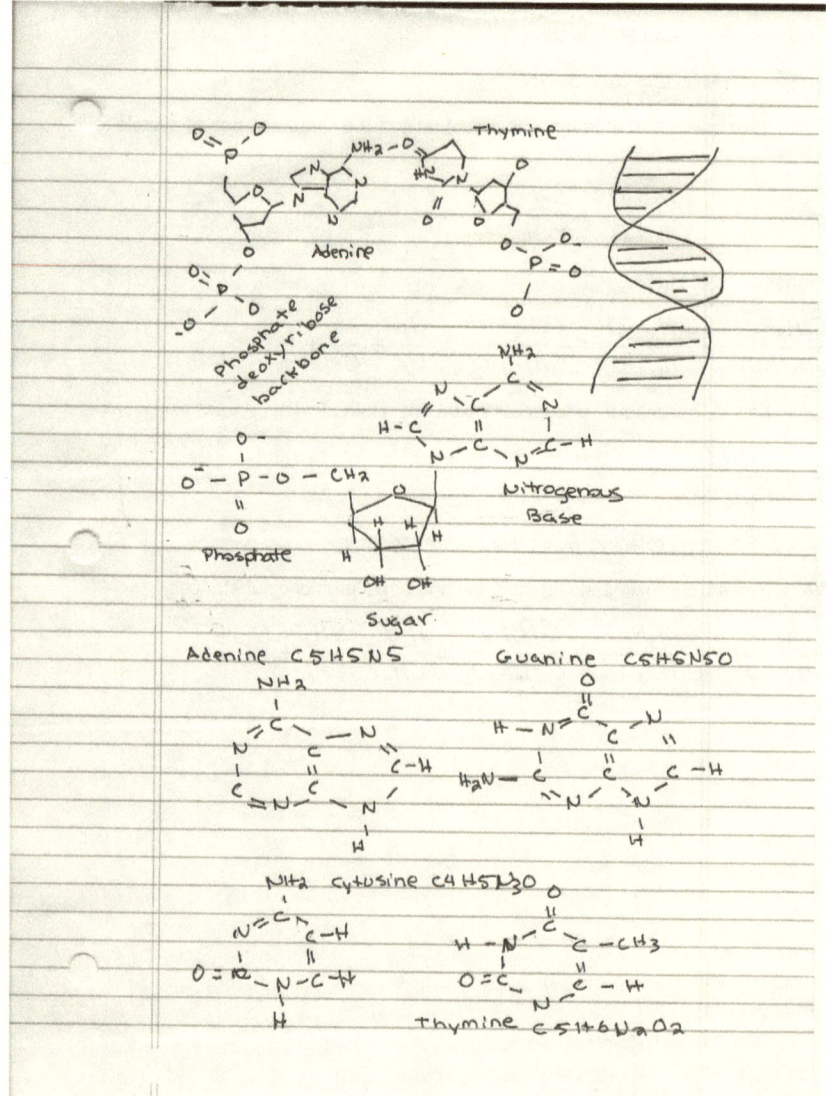

Thymine

Adenine

Phosphate
deoxyribose
backbone

Nitrogenous
Base

Phosphate

Sugar

Adenine C5H5N5   Guanine C5H6N5O

Cytusine C4H5N3O

Thymine C5H6N2O2

Male and Female in AI

We consider the female sex hormone estradiol (estrogen , E):

$$C_{18}H_{24}O_2 = 272.38 g/mol$$

And the male sex hormone testosterone (T):

$$C_{19}H_{28}O_2 = 288.42 g/mol$$

And, cholesterol (Ch) from which both are made:

$$C_{27}H_{46}O = 386.65 g/mol$$

And notice,...

$$\frac{Ch + T}{E} = 2.5$$

And we consider the semiconductor materials used to make AI:

$$\frac{Ge}{Si} = 2.6$$

And write,...

$$\frac{Ch + T}{E} = \frac{Ge}{Si}$$

$$T = \frac{Ge}{Si}E - Ch \qquad E = \frac{Si}{Ge}(T + Ch)$$

$$T\left(1 - \frac{Si}{Ge}\right) + E\left(1 - \frac{Ge}{Si}\right) = Ch\left(\frac{Si}{Ge} - 1\right)$$

We notice that the masculine (T) is in inverse relation to the feminine (E), but that the two add up to on whole (Ch) in that the masculine has coefficient 1-Si/Ge and the feminine has coefficient 1-Ge/Si. This expresses the inverse relationships between man and woman.

Book 3: Asymmetry in Artificial Intelligence

The Asymmetry

The primary elements of artificial intelligence (AI) used to make diodes and transistors, silicon (Si) and germanium (Ge) doped with boron (B) and phosphorus (P) or gallium (Ga) and arsenic (As) have an asymmetry due to boron. Silicon and germanium are in group 14 like carbon (C) and as such have 4 valence electrons. Thus to have positive type silicon and germanium, they need doping agents from group 13 (three valence electrons) like boron and gallium, and to have negative type silicon and germanium they need doping agents from group 15 like phosphorus and arsenic. But where gallium and arsenic are in the same period as germanium, boron is in a different period than silicon (period 2) while phosphorus is not (group 3). Thus aluminum (Al) is in boron's place. This results in an interesting equation.

$$\frac{Si(As - Ga) + Ge(P - Al)}{SiGe} = \frac{2B}{Ge + Si}$$

The differential across germanium crossed with silicon plus the differential across silicon crossed with germanium normalized by the product between silicon and germanium is equal to the boron divided by the average between the germanium and the silicon. The equation has nearly 100% accuracy:

$$\frac{28.09(74.92 - 69.72) + 72.61(30.97 - 26.98)}{(28.09)(72.61)} = \frac{2(10.81)}{(72.61 + 28.09)}$$

$$0.213658912 = 0.21469712$$

$$\frac{0.213658912}{0.21469712} = 0.995$$

99.5%

To illustrate this asymmetry here is that section of the periodic table:

	13	14	15
Period 2	B		
Period 3		Si	P
Period 4	Ga	Ge	As

This I believe will open up the door to a lot in the periodic table of the elements because:

$$\left| \begin{pmatrix} 0 & 0 & Si \\ As & Ga & 0 \\ 1 & 1 & 0 \end{pmatrix} \right| = Si(As - Ga)$$

$$\left| \begin{pmatrix} 0 & 0 & Ge \\ P & Al & 0 \\ 1 & 1 & 0 \end{pmatrix} \right| = Ge(P - Al)$$

But,..

$$\frac{2B}{Ge + Si} = \frac{B}{\frac{Ge + Si}{2}} \quad \text{And,}$$

$$\frac{Ge + Si}{2} = average(Ge, Si)$$

But,

$$average(f) = \frac{1}{b - a} \int_a^b f(x) dx$$

And,

$$\int_S (\nabla \times \vec{u}) \cdot d\vec{S} = \oint_C \vec{u} \cdot d\vec{r}$$

Where,

$$\nabla \times \vec{u} = \left| \begin{pmatrix} \vec{i} & \vec{j} & \vec{k} \\ \frac{\partial}{\partial x} & \frac{\partial}{\partial y} & \frac{\partial}{\partial z} \\ u_1 & u_2 & u_3 \end{pmatrix} \right|$$

But

$$\frac{Si(As - Ga) + Ge(P - Al)}{SiGe} = \frac{2B}{Ge + Si}$$

Can be written

$$\frac{Si}{B}(As - Ga) + \frac{Ge}{B}(P - Al) = \frac{2SiGe}{Si + Ge}$$

But

$$\frac{2SiGe}{Si + Ge}$$

Is the harmonic mean between Si and Ge. However, where we had

$$average(f) = \frac{1}{b - a}\int_{a}^{b} f(x)dx$$

We have

$$harmonic(f) = \frac{1}{\frac{1}{b-a}\int_{a}^{b} f(x)^{-1}dx}$$

As it would turn out

$$\frac{Si}{B}(As - Ga) = \frac{1}{3}H$$

$$\frac{Ge}{B}(P - Al) = \frac{2}{3}H$$

Where

$$H = \frac{2SiGe}{Si + Ge}$$

We have

$$\bar{f} = \frac{1}{b-a}\int_a^b f(x)dx$$

Since we want f(x) such that

$$\frac{1}{Ge-Si}\int_{Si}^{Ge} f(x)dx = \frac{Si+Ge}{2}$$

Then,

$$f(x) = x$$

But,

$$harmonic(f) = \frac{1}{\frac{1}{b-a}\int_a^b f(x)^{-1}dx}$$

$$\frac{Si}{B}(As-Ga) + \frac{Ge}{B}(P-Al) = \frac{Ge-Si}{\int_{Si}^{Ge}\frac{dx}{x}}$$

But,

$$\frac{2SiGe}{Si+Ge} \approx \frac{Si+Ge}{2}$$

$$\int_0^1\int_0^1\left[\frac{Si}{B}(As-Ga) + \frac{Ge}{B}(P-Al)\right]dxdy \approx \frac{1}{Ge-Si}\int_{Si}^{Ge} xdx$$

$$(Ge-Si)\left[\frac{Si}{B}(As-Ga) + \frac{Ge}{B}(P-Al)\right] \approx \int_{Si}^{Ge} xdx$$

$$(72.61-28.09)\left[\frac{28.09}{10.81}(74.92-69.72) + \frac{69.72}{10.81}(30.97-26.98)\right] = \frac{1}{2}(72.61^2-28.09^2)$$

$$1746.5 \approx 2241.5$$

80%

I propose the following two matrices:

$$\left|\begin{pmatrix} \vec{i} & \vec{j} & \vec{k} \\ \frac{\partial}{\partial x} & \frac{\partial}{\partial y} & \frac{\partial}{\partial z} \\ 0 & \frac{Si}{B}(Ga)z & \frac{Si}{B}(As)y \end{pmatrix}\right| = \frac{Si}{B}(As - Ga)\vec{i}$$

$$\left|\begin{pmatrix} \vec{i} & \vec{j} & \vec{k} \\ \frac{\partial}{\partial x} & \frac{\partial}{\partial y} & \frac{\partial}{\partial z} \\ \frac{Ge}{B}(Al)z & 0 & \frac{Ge}{B}(P)x \end{pmatrix}\right| = \frac{Ge}{B}(P - Al)\vec{j}$$

And we have said:

$$\frac{2SiGe}{Si + Ge} \approx \frac{Si + Ge}{2}$$

And,

$$\frac{Si}{B}(As - Ga) = \frac{1}{3}H$$

$$\frac{Ge}{B}(P - Al) = \frac{2}{3}H$$

Where

$$H = \frac{2SiGe}{Si + Ge}$$

Thus if

$$\nabla \times \vec{u} = \frac{Si}{B}(As - Ga)\vec{i}$$

$$\nabla \times \vec{u} = \frac{Ge}{B}(P - Al)\vec{j}$$

$$\vec{dS} = dxdy$$

Then we have the following two integrals:

$$\int_0^1\int_0^1 \frac{Si}{B}(As-Ga)dxdy = \frac{1}{3}\frac{1}{Ge-Si}\int_{Si}^{Ge} xdx$$

$$\int_0^1\int_0^1 \frac{Ge}{B}(P-Al)dxdy = \frac{2}{3}\frac{1}{Ge-Si}\int_{Si}^{Ge} xdx$$

These are very accurate:

$$\frac{28.09}{10.81}(74.92-69.72) = \frac{1}{3}\frac{2(28.09)(72.61)}{28.09+72.61}$$

$$2.5985(5.2) = \frac{1}{3}\frac{4079.2298}{100.7}$$

$$13.5122 = 13.5029$$

99%

$$\frac{72.61}{10.81}(30.97-26.98) = \frac{2}{3}\frac{2(28.09)(72.61)}{28.09+72.61}$$

$$6.7169(3.99) = \frac{2}{3}\frac{4079.2298}{100.7}$$

$$26.8004 = 27.006$$

99%

These are the sets of planes:

$$y = \frac{Si}{B}(Ga)z = 181z, \ z = \frac{Si}{B}(As)y = 195y$$

$$x = \frac{Ge}{B}(Al)z = 181z, \ z = \frac{Ge}{B}(P)x = 208x$$

See the visualizations on the next page,...

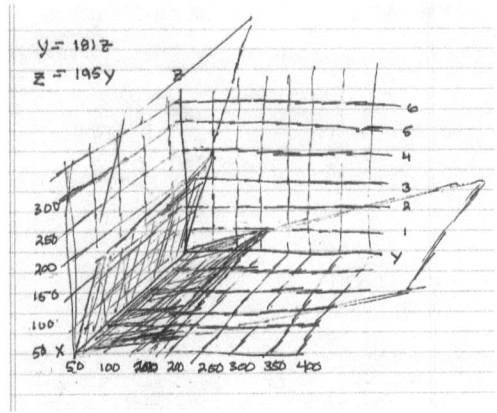

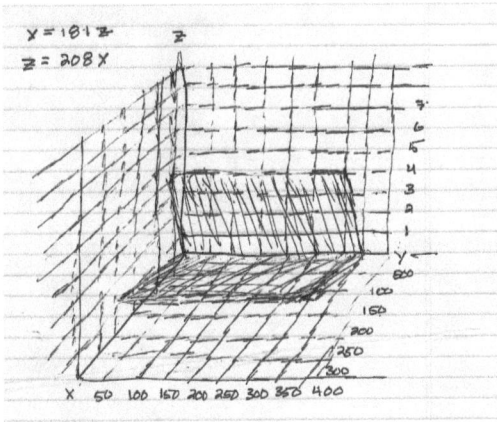

When we say we want f(x) such that

$$\frac{1}{Ge - Si} \int_{Si}^{Ge} f(x)dx = \frac{Si + Ge}{2}$$

It is true but the equations are only 80% accurate so we want f(x) such that

2241.5x=1746.5 or x=0.8=8/10 so that our equations become

$$\int_0^1 \int_0^1 \left[\frac{Si}{B}(As - Ga) + \frac{Ge}{B}(P - Al) \right] dxdy \approx \frac{1}{Ge - Si} \frac{4}{5} \int_{Si}^{Ge} xdx$$

And

$$\int_0^1 \int_0^1 \frac{Si}{B}(As - Ga)dxdy = \frac{1}{4} \frac{1}{Ge - Si} \int_{Si}^{Ge} xdx$$

$$\int_0^1 \int_0^1 \frac{Ge}{B}(P - Al)dxdy = \frac{1}{2} \frac{1}{Ge - Si} \int_{Si}^{Ge} xdx$$

That is for the first equation

$$f(x) = \frac{4}{5}x$$

This gives

$$H\bar{f} = \frac{1}{\frac{1}{72.61 - 28.09}\int_{28.09}^{72.61} \frac{5}{4} \frac{dx}{x}} = \frac{1}{0.002246\frac{5}{4}ln\frac{Ge}{Si}} = 37.6$$

Where $H\bar{f}$ means the harmonic mean of f. Does this approximately equal

$$\frac{2SiGe}{Si + Ge}?$$

$$\frac{2(28.09)(72.61)}{28.09 + 72.61} = 40.5$$

It does.

Line Integrals

We can formulate this a little differently:

$$\vec{u}_1 = \frac{Si}{B}(Ga)z\,\vec{j} + \frac{Si}{B}(As)y\,\vec{k}$$

$$\vec{u}_2 = \frac{Ge}{B}(Al)z\,\vec{i} + \frac{Ge}{B}(P)x\,\vec{k}$$

$$\left(\vec{i}\,\frac{\partial}{\partial x} + \vec{j}\,\frac{\partial}{\partial y} + \vec{k}\,\frac{\partial}{\partial z}\right) \cdot \left(0\vec{i} + \frac{Si}{B}(Ga)z\,\vec{j} + \frac{Si}{B}(As)y\,\vec{k}\right) = 0$$

$$\left(\vec{i}\,\frac{\partial}{\partial x} + \vec{j}\,\frac{\partial}{\partial y} + \vec{k}\,\frac{\partial}{\partial z}\right) \cdot \left(\frac{Ge}{B}(Al)z\,\vec{i} + 0\vec{j} + \frac{Ge}{B}(P)x\,\vec{z}\right) = 0$$

We can write

$$f_1 = \frac{Si}{B}(Ga)z + \frac{Si}{B}(As)y$$

$$f_2 = \frac{Ge}{B}(Al)z + \frac{Ge}{B}(P)x$$

$$\vec{v}_1 = \nabla f_1 = 0\vec{i} + \frac{Si}{B}(As)\vec{j} + \frac{Si}{B}(Ga)\,\vec{k}$$

$$\vec{v}_2 = \nabla f_2 = \frac{Ge}{B}(P)\vec{i} + 0\vec{j} + \frac{Ge}{B}(Al)\,\vec{k}$$

$$\nabla \times \vec{v}_1 \begin{vmatrix} \vec{i} & \vec{j} & \vec{k} \\ \frac{\partial}{\partial x} & \frac{\partial}{\partial y} & \frac{\partial}{\partial z} \\ 0 & \frac{Si}{B}(Ga) & \frac{Si}{B}(As) \end{vmatrix} = 0$$

$$\nabla \times \vec{v}_1 \begin{vmatrix} \vec{i} & \vec{j} & \vec{k} \\ \frac{\partial}{\partial x} & \frac{\partial}{\partial y} & \frac{\partial}{\partial z} \\ \frac{Ge}{B}(P) & 0 & \frac{Ge}{B}(Al) \end{vmatrix} = 0$$

Both are conservative vector fields.

If we define

$$\vec{r}_1 = y\vec{j} + z\vec{k}$$

$$\vec{r}_2 = x\vec{i} + z\vec{k}$$

Then

$$\vec{v}_1 \cdot \vec{r}_1 = \left(\frac{Si}{B}(As)\vec{j} + \frac{Si}{B}(Ga)\vec{k}\right) \cdot \left(y\vec{j} + z\vec{k}\right) = \frac{Si}{B}(As)y + \frac{Si}{B}(Ga)z$$

$$\vec{v}_2 \cdot \vec{r}_2 = \left(\frac{Ge}{B}(P)\vec{i} + \frac{Ge}{B}(Al)\vec{k}\right) \cdot \left(x\vec{i} + z\vec{k}\right) = \frac{Ge}{B}(P)x + \frac{Ge}{B}(Al)z$$

$$\frac{\partial \vec{r}_1}{\partial x} = 0, \frac{\partial \vec{r}_1}{\partial y} = 1, \frac{\partial \vec{r}_1}{\partial z} = 1$$

$$\frac{\partial \vec{r}_2}{\partial x} = 1, \frac{\partial \vec{r}_2}{\partial y} = 0, \frac{\partial \vec{r}_2}{\partial z} = 1$$

$$\frac{d\vec{r}_1}{d\theta} = -R\sin\theta$$

$$\frac{d\vec{r}_2}{d\theta} = R\cos\theta$$

$$-R\frac{Si}{B}(As)\int_0^{\frac{\pi}{2}} \sin\theta d\theta + R(\frac{Si}{B}(Ga)\int_0^{\frac{\pi}{2}} \cos\theta d\theta = R\frac{Si}{B}(As - Ga)$$

For the other integral

$$-R\frac{Ge}{B}(P)\int_0^{\frac{\pi}{2}} \sin\theta d\theta + R\frac{Ge}{B}(Al)\int_0^{\frac{\pi}{2}} \cos\theta d\theta = R\frac{Ge}{B}(P - Al)$$

This is treating

$$\vec{v}_1 = \nabla f_1 = 0\vec{i} + \frac{Si}{B}(As)\vec{j} + \frac{Si}{B}(Ga)\vec{k}$$

And,

$$\vec{v}_2 = \nabla f_2 = \frac{Ge}{B}(P)\vec{i} + 0\vec{j} + \frac{Ge}{B}(Al)\vec{k}$$

As vector fields, and $\vec{r}_1 = y\vec{j} + z\vec{k}$ and $\vec{r}_2 = x\vec{i} + z\vec{k}$ the paths (circular).

Surface Integrals

It is easier to use the divergence theorem and find the surface integral using a volume integral. Namely, that

$$\oint_S \vec{u} \cdot d\vec{S} = \int_V \nabla \cdot \vec{u} \, dV$$

$$\vec{u} = \frac{Si}{B}(As)x\vec{i} + \frac{Si}{B}(Ga)y\vec{j}$$

$$\nabla \cdot \vec{u} = \frac{Si}{B}(As + Ga)$$

$$dV = r^2 \sin\theta \, dr \, d\theta \, d\phi$$

$$\oint_S \vec{u} \cdot d\vec{S} = \frac{Si}{B}(As + Ga)\int_0^{2\pi}\int_0^{\pi}\int_0^{R} r^2 \sin\theta \, dr \, d\theta \, d\phi = \frac{4}{3}\pi\frac{Si}{B}(As + Ga)R^3$$

$$\oint_S \vec{u} \cdot d\vec{S} = \frac{4}{3}\pi\frac{Ge}{B}(P + Al)R^3$$

These are the flux of vector fields

$$\vec{u} = \frac{Si}{B}(As)x\vec{i} + \frac{Si}{B}(Ga)y\vec{j}$$

$$\vec{u} = \frac{Ge}{B}(P)x\vec{i} + \frac{Ge}{B}(Al)y\vec{j}$$

For a sphere.

Conservative Vector Fields

For a conservative vector field we must have for a vector field $\vec{u}$

1. $\vec{u} = \nabla \phi$
2. $\nabla \times \vec{u} = 0$
3. $\int_C \vec{u} \cdot d\vec{r}$ is path independent
4. $\oint_C \vec{u} \cdot d\vec{r} = 0$

We have

$$\vec{u} = \frac{Si}{B}(Ga)x\vec{i} + \frac{Si}{B}(As)y\vec{j}$$

$$\frac{\partial \phi}{\partial x} = \frac{Si}{B}(Ga)x$$

$$\frac{\partial \phi}{\partial y} = \frac{Si}{B}(As)y$$

$$\phi = \int \frac{Si}{B}(Ga)x\,dx = \frac{Si}{B}(Ga)\frac{x^2}{2} + f(y)$$

$$\frac{\partial \phi}{\partial y} = f'(y) = \frac{Si}{B}(As)y$$

$$\int f'(y)\,dy = \int \frac{Si}{B}(As)y\,dy$$

$$f(y) = \frac{Si}{B}(As)\frac{y^2}{2} + C$$

$$\phi = \frac{Si}{B}(Ga)\frac{x^2}{2} + \frac{Si}{B}(As)\frac{y^2}{2} + C$$

$$\phi = \frac{Ge}{B}(P)\frac{x^2}{2} + \frac{Ge}{B}(Al)\frac{y^2}{2} + C$$

Phi (ϕ) is a potential.

Input:

$$z(x, y) = \frac{1}{2}(208\,x^2) + \frac{1}{2}(181\,y^2)$$

Result:

$$z(x, y) = 104\,x^2 + \frac{181\,y^2}{2}$$

Geometric figure:

elliptic paraboloid

3D plot:

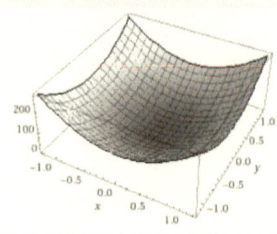

Input:

$$z(x, y) = \frac{181}{2}\,x^2 + \frac{195}{2}\,y^2$$

Geometric figure:

elliptic paraboloid

3D plot:

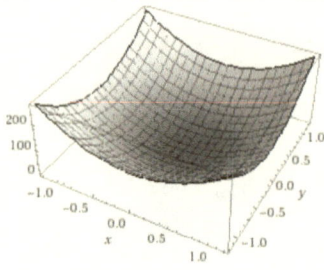

Density

The mass of a sphere with density ρ_0 at its center and ρ_1 at its surface that varies linearly from its center is given by

$$M = \iiint_V \rho(x, y, z)dxdydz$$

Where

$$\rho(r) = \rho_0 + (\rho_1 - \rho_0)\frac{r}{R}$$

In spherical coordinates the change of variables results in following triple integral:

$$M = \int_0^{2\pi} \int_0^{\pi} \int_0^{R} \left[\rho_0 + (\rho_1 - \rho_0)\frac{r}{R}\right] r^2 sin\theta dr d\theta d\phi$$

$$M = \int_0^{2\pi} d\phi \int_0^{\pi} sin\theta d\theta \int_0^{R} \left[\rho_0 + (\rho_1 - \rho_0)\frac{r}{R}\right] r^2 dr$$

$$M = (2\pi)(-cos(\pi) + cos(0)) \int_0^{R} \left[\rho_0 + (\rho_1 - \rho_0)\frac{r}{R}\right] r^2 dr$$

$$M = 4\pi \int_0^{R} \left[\rho_0 + (\rho_1 - \rho_0)\frac{r}{R}\right] r^2 dr$$

In general for a scalar field that depends only on the distance from the origin radially

$$\int_V f dV = 4\pi \int_0^{R} f(r) r^2 dr$$

$$M = \frac{4}{3}\pi R^3 (\frac{1}{4}\rho_0 + \frac{3}{4}\rho_1)$$

Dividing out the volume we have the average density:

$$\bar{\rho} = \frac{1}{4}\rho_0 + \frac{3}{4}\rho_1$$

Silicon is 2.33 grams per cubic centimeter and germanium is 3.323 grams for cubic centimeter. If the density at the center is silicon and germanium at the surface we have the average density is 4.57475 grams per cubic centimeter which is titanium (Ti) at 4.5 grams per cubic centimeter and if it is germanium at the center and silicon at the surface then we have 3.07825 grams per cubic centimeter which is scandium (Sc) at 3.00 grams per cubic centimeter. Both titanium and scandium are light, strong metals that are alloyed with aluminum.

Radius, Density, And Molar Mass

The golden ratio and the golden ratio conjugate are the solution of the quadratic

$$\left(\frac{a}{b}\right)^2 - \frac{a}{b} - 1 = 0 \text{ that meets the conditions } \frac{a}{b} = \frac{b}{c} \text{ and } a=b+c$$

Where $\Phi = \dfrac{a}{b}$ and $\phi = \dfrac{\sqrt{5}-1}{2}$, $\phi = \dfrac{1}{\Phi}$.

We guess that artificial intelligence (AI) has the golden ratio, or its conjugate in its means geometric, harmonic, and arithmetic by molar mass by taking these means between doping agents phosphorus (P) and boron (B) divided by semiconductor material silicon (Si) :

$$\frac{\sqrt{PB}}{Si} = \frac{\sqrt{(30.97)(10.81)}}{28.09} = 0.65$$

$$\frac{2PB}{P+B}\frac{1}{Si} = \frac{2(30.97)(10.81)}{30.97+10.81}\frac{1}{28.09} = 0.57$$

$$\frac{0.65+0.57}{2} = 0.61 \approx \phi$$

Which can be written

$$\frac{\sqrt{PB}(P+B)+2PB}{2(P+B)Si} \approx \phi$$

We now want to write out the equations for atomic radius, density, and molar mass as these are the components upon which the properties of the elements should rely.

P_R	Radius Phosphorus	100 pm	
B_R	Radius Boron	85 pm	
Si_R	Radius Silicon	110 pm	
Ga_R	Radius Gallium	130 pm	
As_R	Radius Arsenic	115 pm	
Ge_R	Radius Germanium	125 pm	
P_M	Molar Mas Phosphorus	30.97 g/mol	
B_M	Molar Mass Boron	10.81 g/mol	
Si_M	Molar Mass Silicon	28.09 g/mol	

You will find:

$$\left[\frac{Ga_R}{Ge_R} + \frac{As_R}{Ge_R} + 1\right] \approx \pi \quad \text{and} \quad \left[\frac{P_R}{Si_R} + \frac{B_R}{Si_R}\right] \approx \Phi$$

Or,...

$$\frac{P_R + B_R}{Si_R} \approx \Phi \quad \text{Or,....} \quad \frac{Si_R}{P_R + B_R} \approx \phi$$

We now subscript the elements wlth M for molar mass using an earlier equation:

$$\frac{\sqrt{P_M B_M}(P_M + B_M) + 2P_M B_M}{2(P_M + B_M)Si_M} \approx \phi$$

Which can be written,...

$$\frac{\sqrt{P_M B_M} + \frac{2P_M B_M}{(P_M + B_M)}}{2Si_M} \approx \phi$$

Which yields,...

$$\frac{2Si_R}{P_R + B_R}Si_M \approx \sqrt{P_M B_M} + \frac{2P_M B_M}{P_M + B_M}$$

The Author

www.ingramcontent.com/pod-product-compliance
Lightning Source LLC
Chambersburg PA
CBHW031122180526
45160CB00005B/59/J